MY FIRST MATHEMATICAL STROKES

An exciting activity book for little explorers in kindergarten and preschool.

What awaits you on this Jurassic journey?

Exciting strokes: Travel along straight and curved lines, discover geometric shapes and figures, and practice numbers 1-10.

Friendly dinosaurs: Join friendly dinosaurs on each page and learn with them in a fun way.

Fine motor skills to the maximum: Develop your hand coordination and precision while exploring the mathematical world.

Mathematics for little adventurers: Immerse yourself in the first mathematical notions in a creative and full of surprises way.

Are you ready for an unforgettable mathematical adventure?

My First Mathematical Strokes is the perfect tool for:

- Stimulate fine motor skills and visual-motor coordination.
- Introduce the first mathematical notions in a playful way.
- Encourage creativity and imagination.
- Strengthen concentration and attention.
- Awaken a taste for mathematics.

Embark on this Jurassic journey and discover a world of fun-filled learning!

Mariledys Tovar

MATHEMATICAL ADVENTURES WITH DINOSAURS

Scan to receive free information and resources

You can also write to
mariledys@educkidsonline.com

This book belongs to:

N
O
T
E
S

-
-
-
-
-

MATHEMATICAL ADVENTURES
WITH DINOSAURS

ÍNDICE

- **7** — HORIZONTAL STRAIGHT LINES
- **10** — VERTICAL STRAIGHT LINES
- **12** — STRAIGHT DIAGONAL LINES
- **14** — WAVY AND CURVED LINES
- **27** — SPIRAL LINES
- **31** — 2D SHAPES AND FORMS
- **47** — NUMBERS FROM 0 TO 10

MY FIRST MATHEMATICAL DRAWINGS

LET'S START

MATHEMATICAL ADVENTURES
WITH DINOSAURS

Dinosaur Tracing

Draw all horizontal lines

Dinosaur Tracing

Draw all horizontal lines

Dinosaur Tracing

Draw a horizontal line

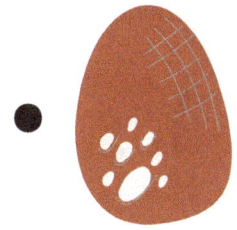

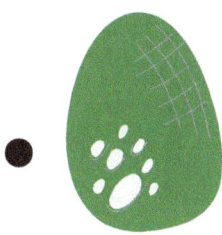

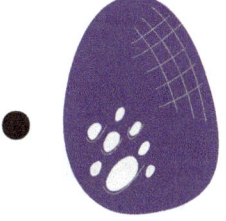

Dinosaur Tracing

Draw the vertical lines.

Dinosaur Tracing

Draw the vertical lines.

Dinosaur Tracing

Draw the diagonal lines

Dinosaur Tracing

Draw the diagonal lines

Dinosaur Tracing

Help these dinosaurs find their way to their leafy snack!

Dinosaur Tracing

Draw the wavy lines

Dinosaur Tracing

Draw the lines

Dinosaur Tracing

Draw the lines

Dinosaur Tracing

Draw the lines

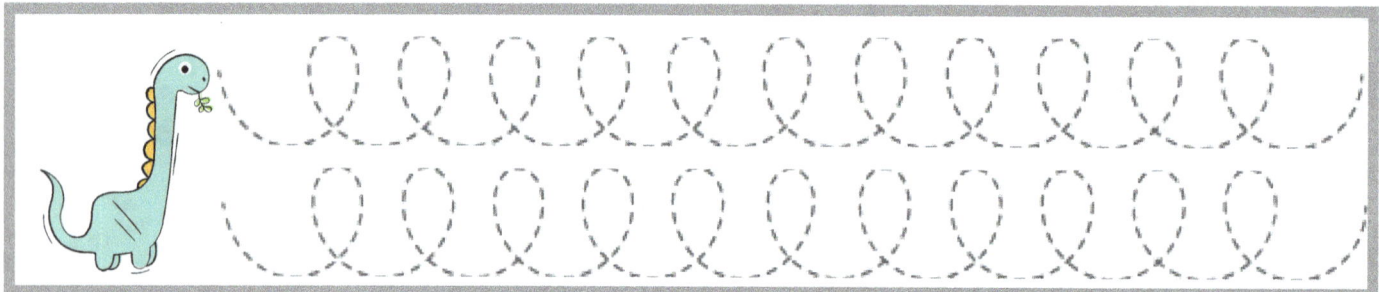

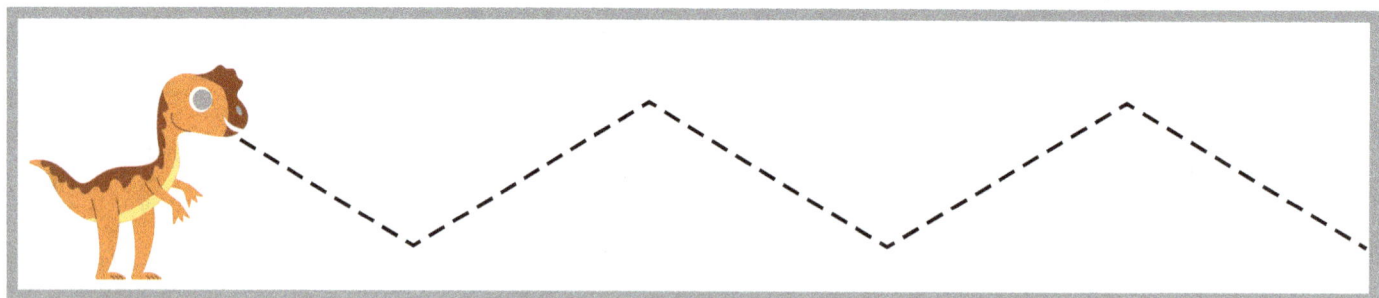

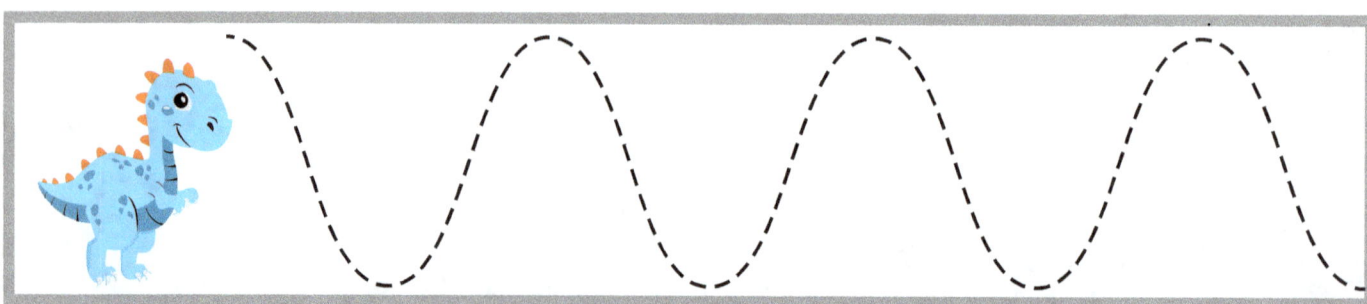

Dinosaur Tracing

Help these dinosaurs follow the footprints!

Dinosaur Tracing

Help the dinosaur find the egg.

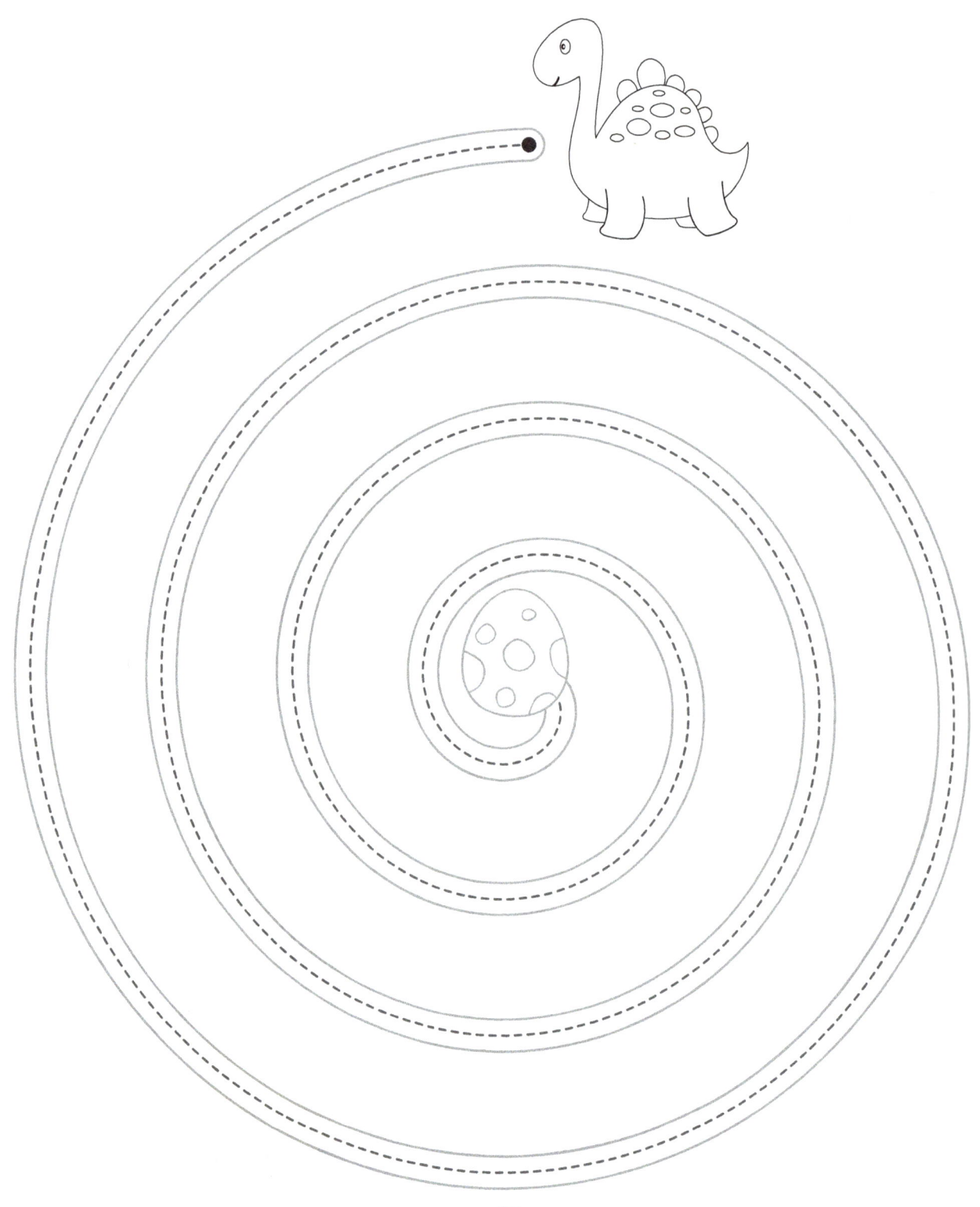

Dinosaur Tracing

Help the dinosaur find the egg.

Dinosaur Tracing

Help the dinosaur find the egg.

2D Shapes

Connect the dots to form the figures.

2D Shapes

Connect the dots to form the figures.

Square

2D Shapes

Connect the dots to form the figures.

Rectangle

2D Shapes

Connect the dots to form the figures.

2D Shapes

Connect the dots to form the figures.

2D Shapes

Connect the dots to form the figures.

2D Shapes

Connect the dots to form the figures.

2D Shapes

Connect the dots to form the figures.

2D Shapes

Connect the dots to form the figures.

2D Shapes

Connect the dots to form the figures.

2D Shapes

Connect the dots to form the figures.

2D Shapes

Connect the dots to form the figures.

2D Shapes

Connect the dots to form the figures.

Color　　　Trace　　　Draw

2D Shapes

Connect the dots to form the figures.

2D Shapes

Connect the dots and complete the figure.

2D Shapes

Connect the dots and complete the figure.

Dino Numbers

Trace and write

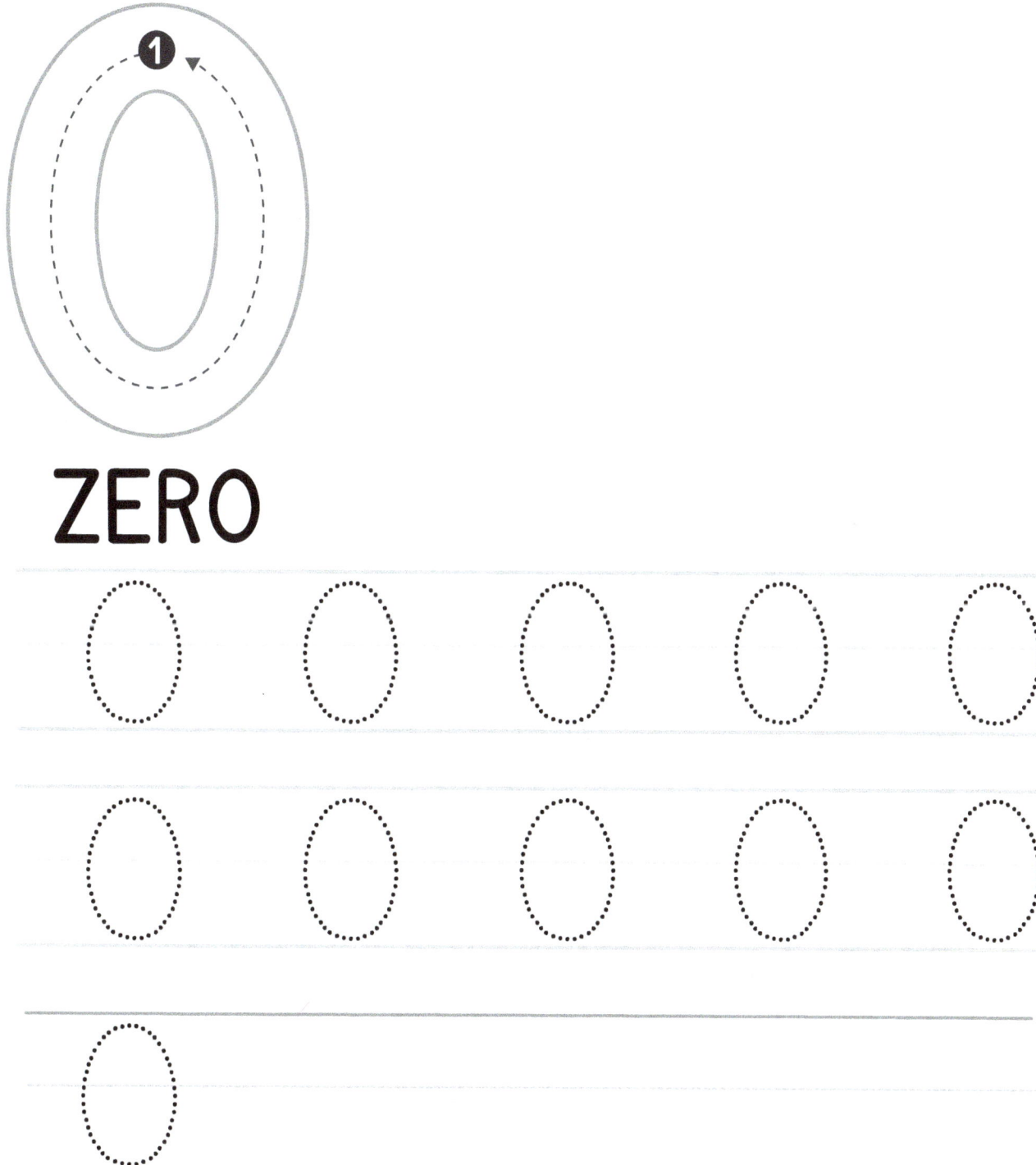

Dino Numbers

Trace and write

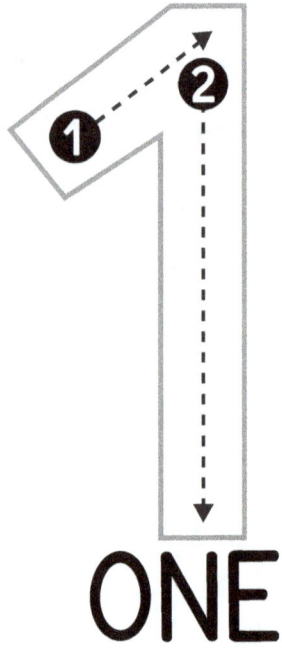

ONE

Dino Numbers

Trace and write

Dino Numbers

Trace and write

THREE

Dino Numbers

Trace and write

Dino Numbers

Trace and write

Dino Numbers

Trace and write

Dino Numbers

Trace and write

SEVEN

Dino Numbers

Trace and write

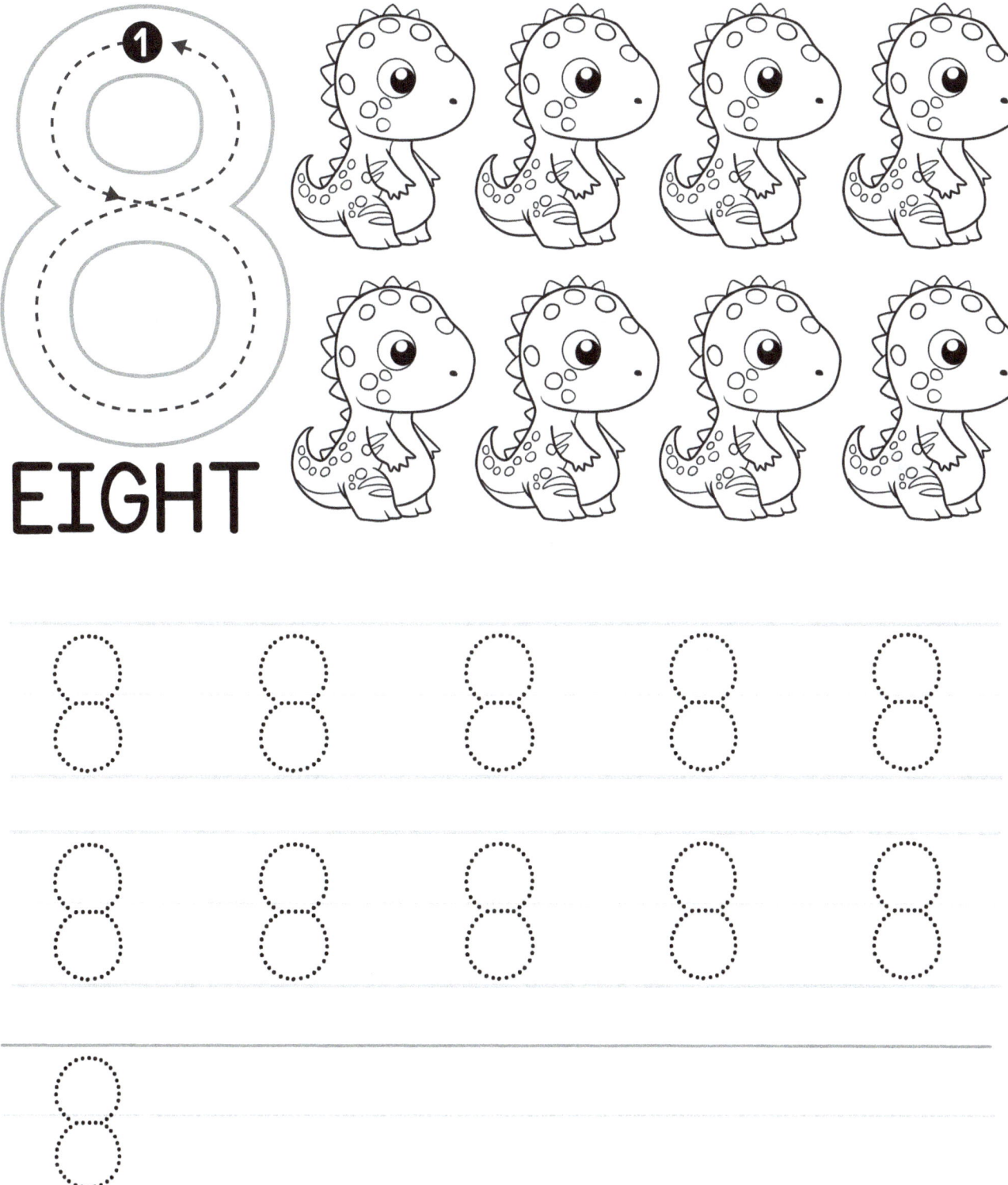

Dino Numbers

Trace and write

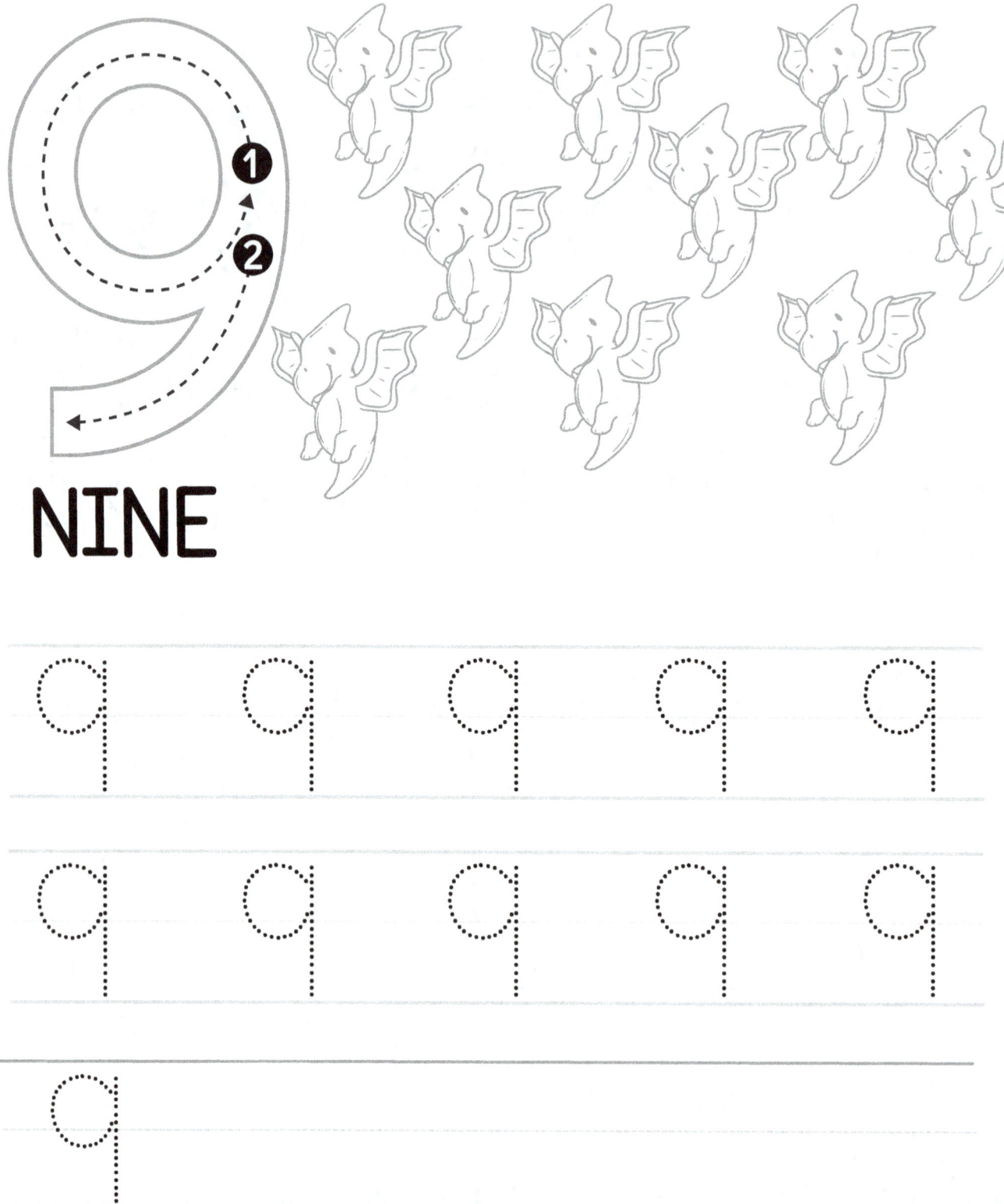

NINE

Dino Numbers

Trace and write

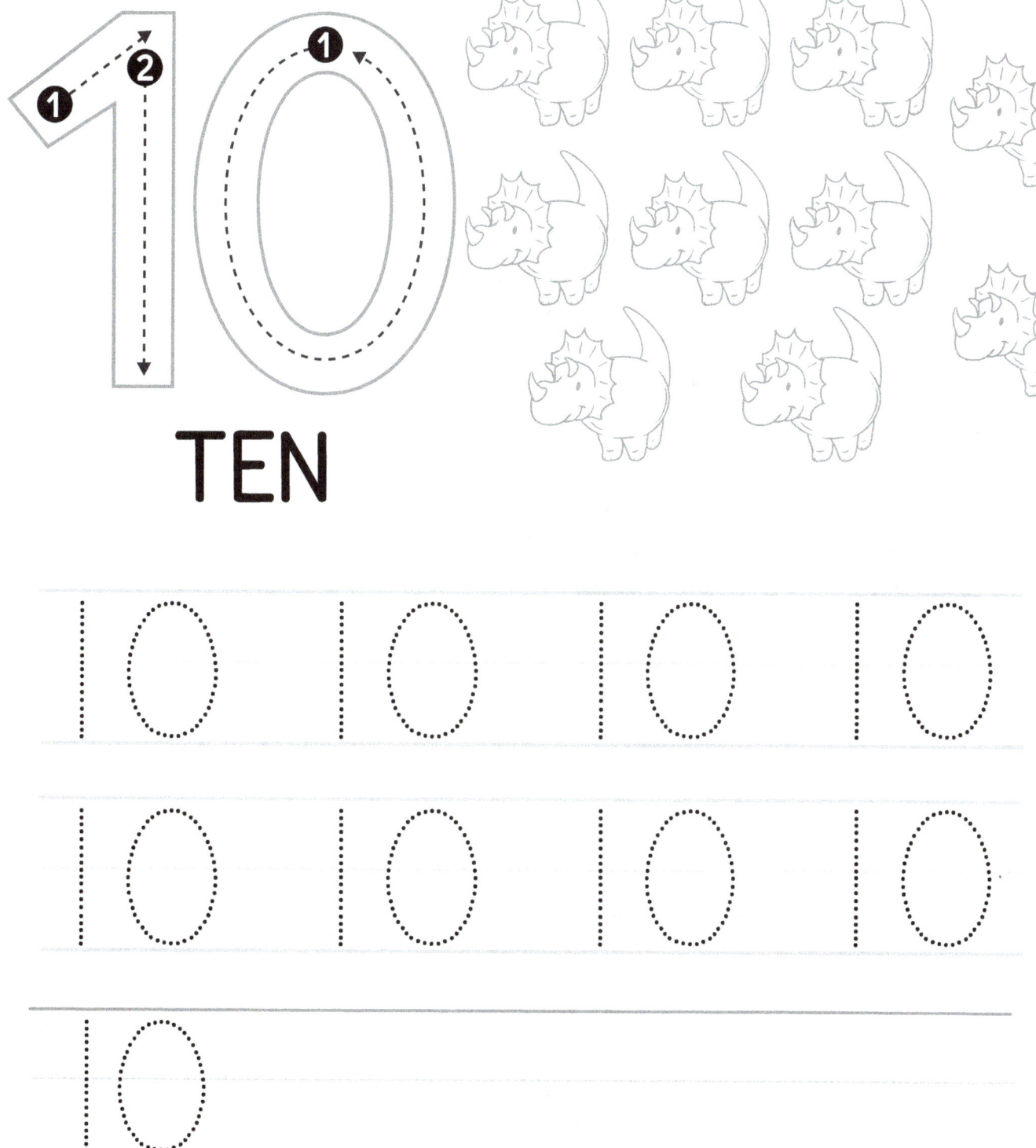

TEN

Dino Numbers

Trace and write

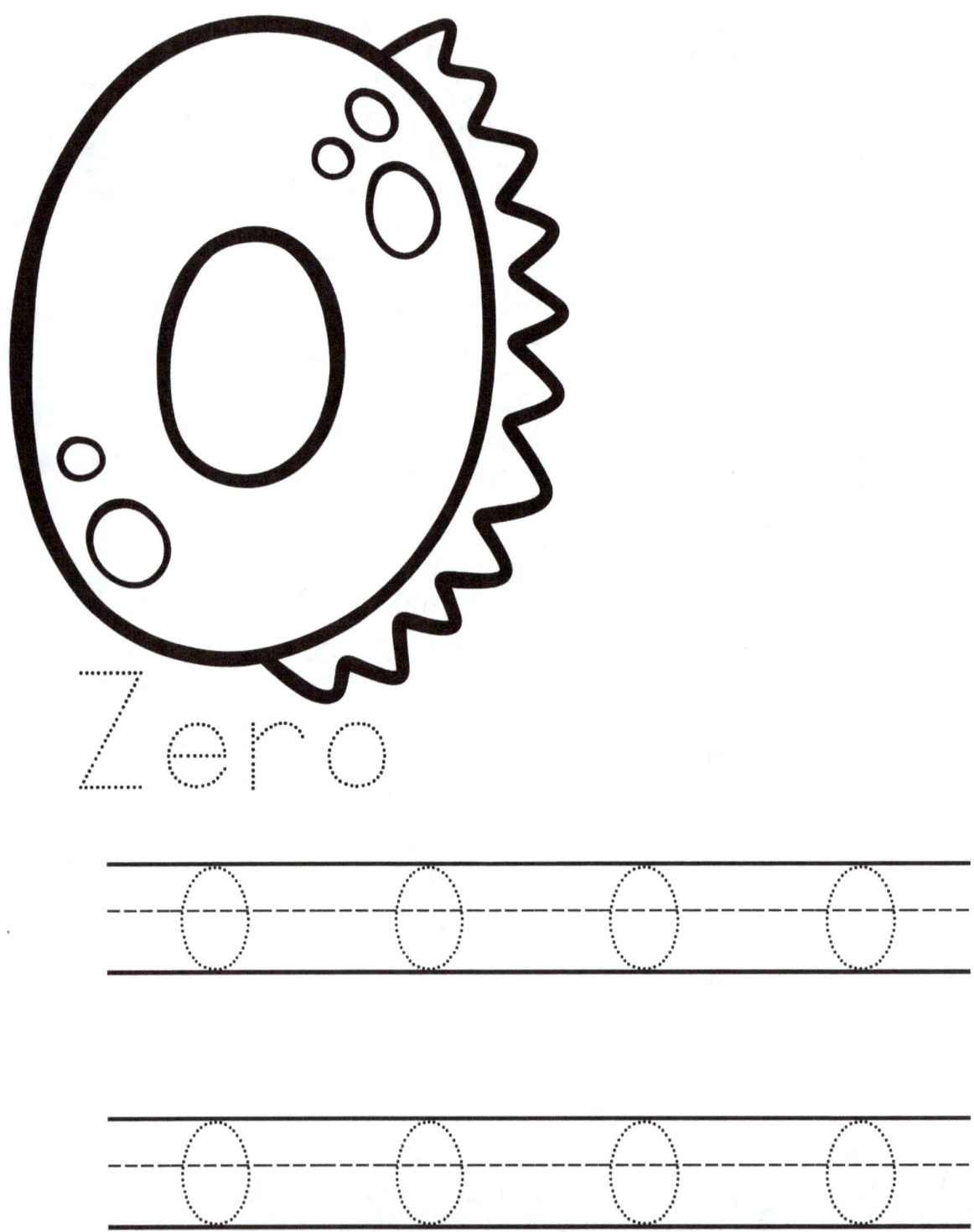

Dino Numbers

Draw, write and color.

Dino Numbers

Draw, write and color.

Dino Numbers

Draw, write and color.

Dino Numbers

Draw, write and color.

Dino Numbers

Draw, write and color.

Dino Numbers

Draw, write and color.

Dino Numbers

Draw, write and color.

Dino Numbers

Draw, write and color.

Eigth

8 8 8 8 8

8 8 8 8 8

Dino Numbers

Draw, write and color.

Dino Numbers

Draw, write and color.

Dino Numbers

Identify, Trace, Write and Draws

READ	WRITE	FIND

TRACE	QUANTITY	

Dino Numbers

Identify, Trace, Write and Draws

READ

WRITE

FIND

TRACE

QUANTITY
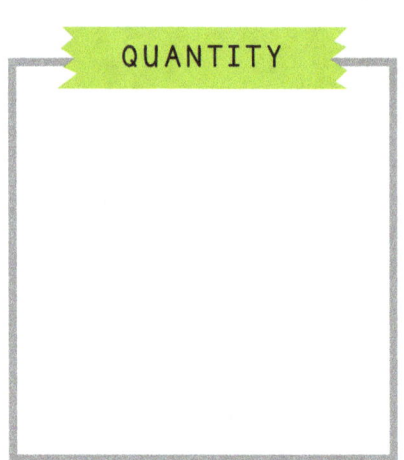

2 2 2 2 2 2 2

2 2 2 2 2 2 2

Dino Numbers

Identify, Trace, Write and Draws

READ

WRITE

FIND

TRACE

QUANTITY
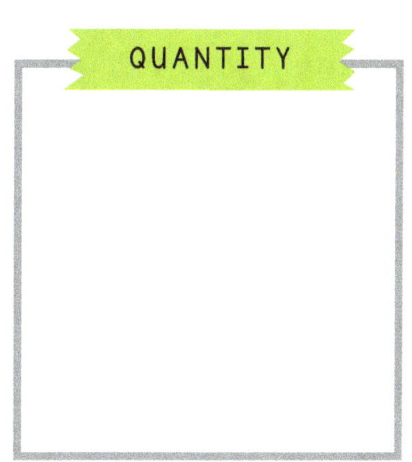

Dino Numbers

Identify, Trace, Write and Draws

READ	WRITE	FIND

TRACE	QUANTITY	

Dino Numbers

Identify, Trace, Write and Draws

READ	WRITE	FIND

TRACE	QUANTITY	
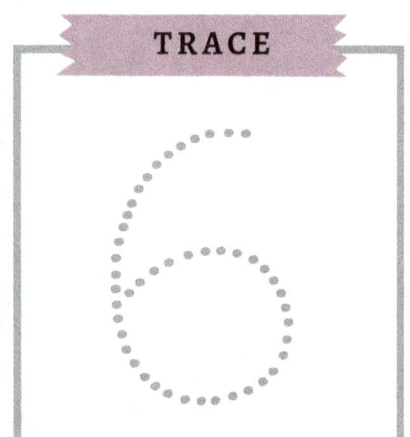	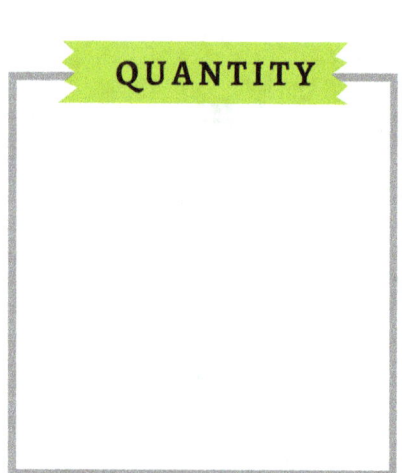	

Dino Numbers

Identify, Trace, Write and Draws

READ	WRITE	FIND

TRACE	QUANTITY	
	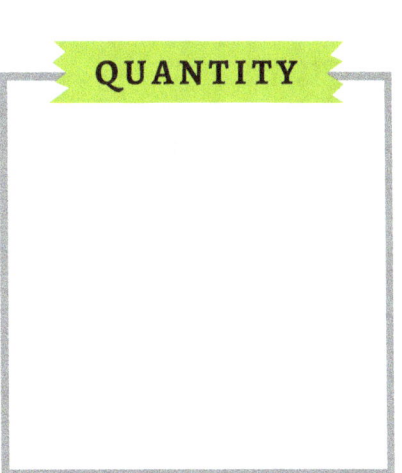	

Dino Numbers

Identify, Trace, Write and Draws

READ	WRITE	FIND

TRACE	QUANTITY	

Dino Numbers

Identify, Trace, Write and Draws

READ	WRITE	FIND

TRACE	QUANTITY
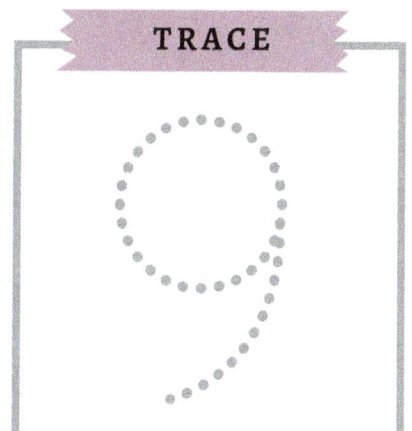	

Dino Numbers

Count the number of objects and circle the correct answer and write it down.

| 5 | 6 | 2 | 1 |

| 5 | 4 | 2 | 3 |

| 1 | 5 | 3 | 2 |

| 7 | 4 | 5 | 9 |

Dino Numbers

Count the number of objects and circle the correct answer and write it down.

9   3   0   6

1   7   6   3

5   9   1   8

3   4   6   8

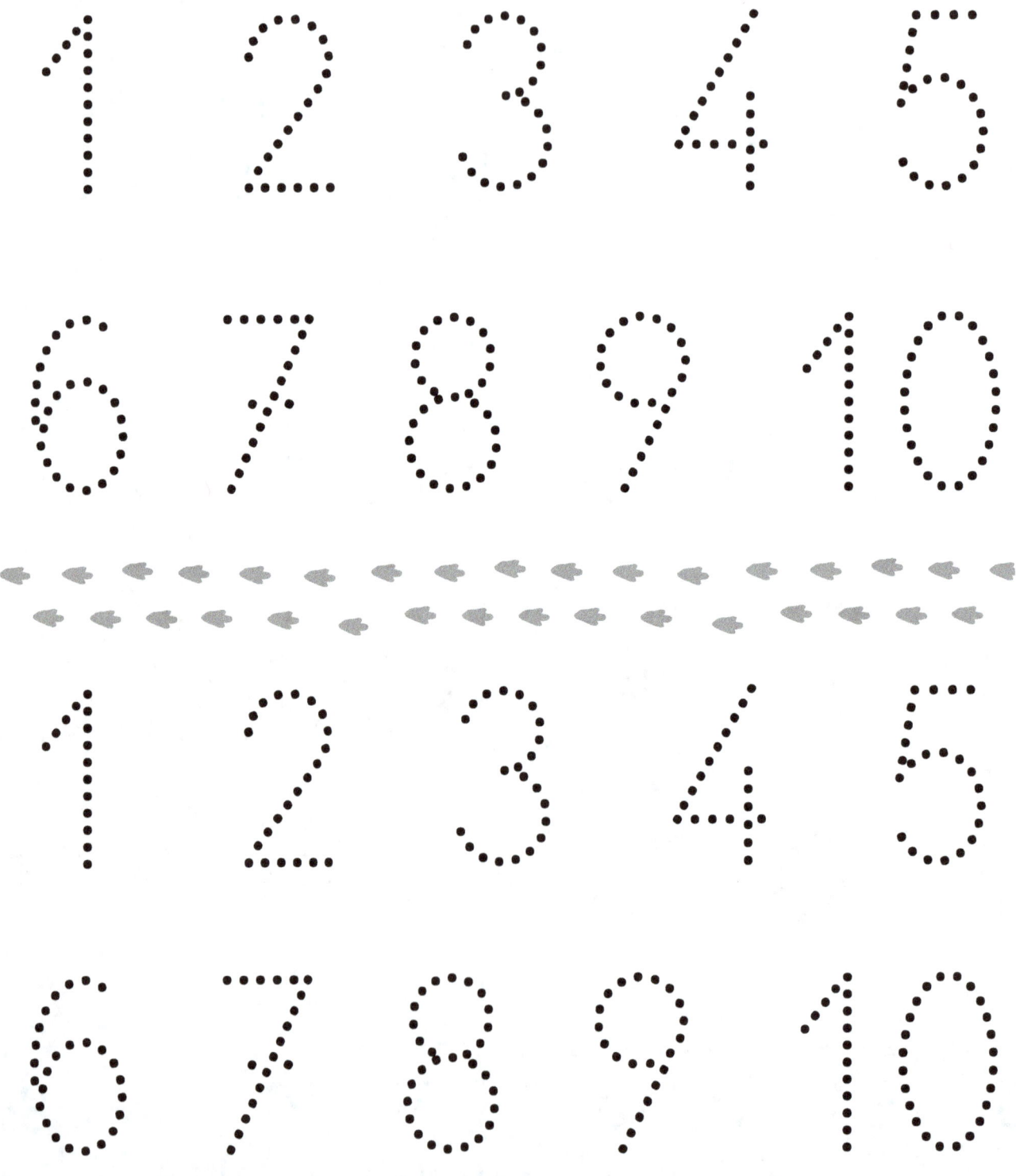

Dino Numbers

Identifies and traces numbers

Dino Numbers

Identifies and traces numbers

0 zero	0 0 0 0 0
1 one	1 1 1 1 1 1
2 two	2 2 2 2 2
3 three	3 3 3 3 3 3
4 four	4 4 4 4 4 4
5 five	5 5 5 5 5 5
6 six	6 6 6 6 6 6
7 seven	7 7 7 7 7 7
8 eight	8 8 8 8 8 8
9 nine	9 9 9 9 9 9
10 ten	10 10 10

Dino Numbers

Write the missing numbers

Dino Numbers

Write the missing numbers

Dino Numbers

Write the missing numbers

Dino Numbers

Write the missing numbers

Dear Readers,

Thank you for embarking on this exciting mathematical logic adventure with me! It has been a pleasure to create this workbook designed especially for curious and creative minds.

The support of readers like you is invaluable to independent authors like me. If you have a moment, I would greatly appreciate it if you could leave a review on Amazon. Your opinions are crucial for more parents, teachers or adults who love early childhood education to find these exercises.

I invite you to visit my Amazon page and by following me you can access the different books I am preparing for our children.

Thank you for contributing to the community of readers and for making it possible for more people to discover this book!

With gratitude, Mariledys

Scan to leave your comment
or visit my Amazon page

www.ingramcontent.com/pod-product-compliance
Lightning Source LLC
Chambersburg PA
CBHW062225220526
45471CB00009B/3349